Dunkirk
Brussels
la Chapelle
NORTH SEA
Dover
Waterloo
Ramillies
GERMANY
Metz
Long. West
Long. East
from London

First Printing
ISBN 9780692713211

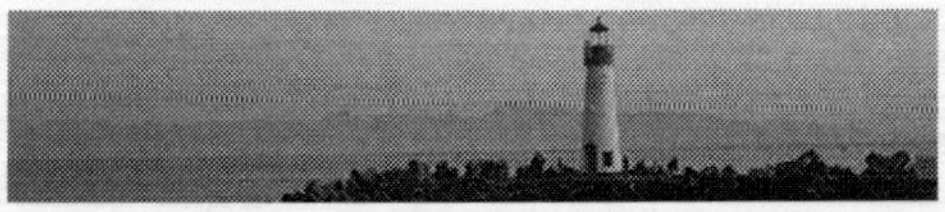

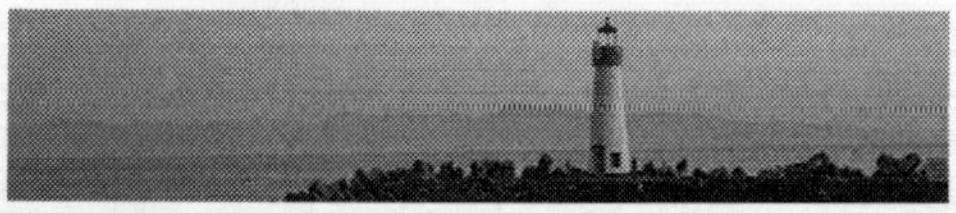

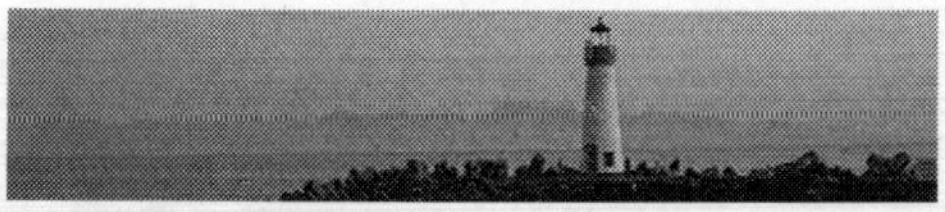

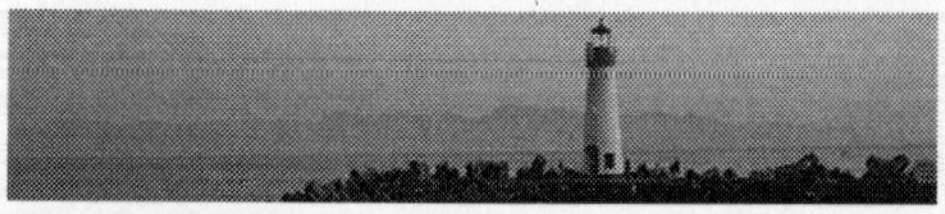

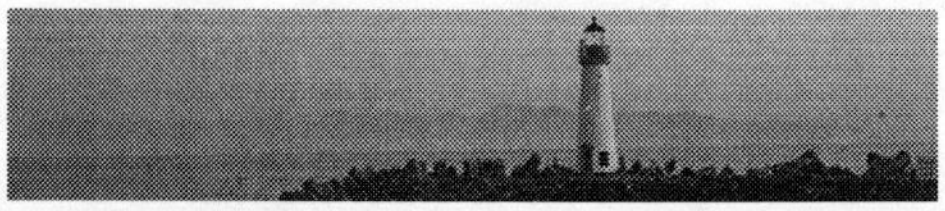

Made in the USA
Monee, IL
04 November 2026

67381751R00095